PROCESS PLANT RELIABILITY

and

MAINTENANCE

for

PACESETTER PERFORMANCE

PROCESS PLANT RELIABILITY

and

MAINTENANCE

for

PACESETTER
PERFORMANCE

by REX KENYON

PennWell®

Copyright © 2004 by
PennWell Corporation
1421 South Sheridan Road
Post Office Box 1260
Tulsa, Oklahoma 74112-6600 USA

800.752.9764
+1.918.831.9421
sales@pennwell.com
www.pennwellbooks.com
www.pennwell.com

Managing Editor: Marla Patterson
Cover Design: Beth Caissie
Book Design: Robin Remaley

Library of Congress Cataloging-in-Publication Data
Available on Request

Kenyon, Rex
Process Plant Reliability and Maintenance for Pacesetter Performance
p.cm.
Includes index.

ISBN 1-59370-024-5

Printed in the United States of America

04 05 06 07 08   5 4 3 2 1

Contents

Preface

This book is written at the request of our clients. They have asked many times, "Is there a book we can read that explains what you are doing to help us? Is there something we could read that pulls it all together? There should be something out there."

My response has always been, "We have not found one yet, but let me draw you a picture of how we see it all fitting together." One picture is our simplified model for routine maintenance work processes included in this book.

This book focuses on maintenance, reliability, work processes, and the appropriate metrics. Regardless of your particular business focus, you need to understand the basics of your business if you are in it for the long haul. If you understand your business well, you can develop a simplified business model for it and chart your path forward.

This book covers very basic concepts. It should not take more than an hour or two to read. I would also recommend that after reading through the entire book quickly, you then study the concepts and think about the model we have presented.

The book is pretty simple, but someone needed to write it. It represents one path torward attaining maintenance and reliability pacesetter performance in the manufacturing arena.

Acknowledgments

Beverly J. Kenyon,
President, Rex Kenyon and Associates, Inc.

Beverly is my wife, mother of our two grown sons, grandmother for our only granddaughter, president of our company, and my supporter. She insisted I write this book. She has been insisting I do this for many years now. She felt it would make a difference for a lot of businesses. I hope this gets our company president off my back!

Dr. Harold S. Haller,
President, Harold S. Haller & Company, Inc.

Dr. Haller is a Deming master. I was fortunate to be a member of a management team when Dr. Haller assisted us in transforming our facility to follow the teachings of Dr. W. Edwards Deming.

Dr. Haller provided us with a copy of his multivariate analytical tool (Multiple Correlation Analysis—HITS-MCA) to solve complex problems at our clients' facilities in an attempt to interest them in becoming users of multivariate analysis.

Dr. Haller's HITS-MCA program and training can be acquired for less than $3,000 per trainee. Typically MCA is used to solve problems worth millions of dollars per year to a manufacturing facility. Problems of this magnitude are almost always solved during the training classes referenced in this paragraph. Dr. Haller is a pretty smart guy, but he does not charge enough for his software and training classes, in my opinion.

How To Use This Book

If you choose to make the suggested improvements on your own, I would recommend that you quickly read through the entire book. I would then recommend that you reread the section on benchmarking (chapter 1) and use that information to determine where your largest gaps exist relative to attaining pacesetter performance. Contact external benchmarking services if you do not know how your performance relates to that of your competitors.

Once you have benchmarking data, reread and follow the guidance in the various chapters that align with your largest performance gaps. These are areas where you have the largest opportunities to reduce costs and increase reliability.

With each gap you attack, make sure you reread the appropriate chapter that addresses the gap as well as the chapter on metrics (chapter 5). Monitoring metrics relative to these gaps will drive the behaviors and work processes that will generate the desired results.

Make sure that you identify the work processes that will deliver the results you desire. Make sure the work processes and metrics are not only necessary, but also that they are sufficient to give you the results you desire.

Once a work process is in place, begin gathering the data for the metrics to monitor performance and reinforce the necessary work processes that will drive your desired results. The work process metrics will be the bellwether indicators for the results you desire. Do not forget to track the results metrics.

Should you go it alone? Sometimes it is necessary. Sure—we could probably help you go faster, and we would do so if you requested, but our services are expensive. Your size of the prize has to be large to get us involved.

Are there any concerns on my part? Not if you follow through. We know that results will come if you follow our process and stay the course. We have guided many organizations and facilities through our process, so we know it works. That is one of the significant benefits we bring to our clients. My only concern is that you might not have the confidence in the process and might not stay the course long enough to see the results. If you have questions or concerns, call us. Our phone number is (909) 288-7027. (If you call, please remember that we are in the Pacific Standard Time.)

Introduction

Fortunately for most people, we are not talking about rocket science in this book. Top performance in any activity requires that those involved do the things they already know should be done. Unfortunately, too many people just do not seem to have the time to do the right things at the right times. We hear comments like:

- "I think it is the *meetings*."

- "It's *meet-o-rama*."

- "I'm the *meeting-meister*."

- "*Meet-o-mania*."

The discussion in meetings is not important. It is the *doing* that is important...whatever *it* is. That is pretty much the case in all we do in life.

Industry now has meetings for everything. A client (manager) told me one time, "You get to do the fun things—making it happen—and I am tied up in the office all day in meetings. I wish we could switch jobs!"

This manager knew he could do what we were doing and he knew how we were doing it—he just did not have time to make it happen. But he knew it would take him longer to do even if he chose to attempt this himself. When he would finish with his meeting marathons, he would return calls to the people who had already left for the day, answer his e-mails, and head for home. Many of our update discussions regarding implementation efforts occurred via his cell phone while he drove to work in the morning or as he drove home at the end of the day. There was no other time in his day—and we were working on survival issues for his business!

We see this as a common thread with many of our clients. Managers literally spend their entire day in meetings talking about doing something rather than actually doing something.

Regarding my comment that this is not rocket science, I do need to clarify that one of the techniques discussed in chapter 4, "Reliability," borders on rocket science. The creator of the methodology I reference will argue that it is not rocket science, just an excellent statistical analysis tool. Rocket science or not, only a couple of people at a facility need to have that level of knowledge. The rest of this information is just about doing the basics we already know we should be doing.

A young engineer (and mother) in Canada agreed on how basic one of the concepts is. After our discussion, she said, "Girl Guides: Fail to plan, plan to fail." That pretty much sums up one of the strategies!

Implementation efforts go very quickly with our involvement, because we have done it before. We know the tricks. We know the potential shortcuts. And we know the pitfalls to avoid. We also have 100% of our day to focus on implementation and to assist with resolving roadblocks. Those who attempt to go it alone will succeed if they stay the course. It just might take a little longer when implementing one of the strategies for the first time.

Top-performing manufacturers (we will use the term *pacesetters*) have higher reliability, higher facility utilization, and less maintenance on their facilities than their competitors. They do less maintenance because they are more reliable. They have higher plant/production line utilization because their facilities are more reliable and they shut down less often to do maintenance. They do less work, and what work they do, they do effectively.

We have seen many examples where a management team tries to reduce maintenance costs on their own. The cycle is to cut maintenance manpower without changing any of the maintenance work processes. The lower manning levels can no longer keep up with the repair workload, since the existing processes have not changed to provide any added effectiveness to the organization. Key equipment starts to break, plant reliability declines, plants start to shut down due to reliability problems, and production falls. Contractors are brought back into the plant to reduce the backlog that has grown since the initial reduction in manpower. Often manpower at this point ends up at levels that are higher than they were prior to the reductions.

The hope of the management team is that they can get promoted for cutting costs (manpower) prior to the point in the cycle where the plant starts to fall apart. The lucky ones do get out in time. The unlucky ones do not, and they suffer the consequences. It is even more devastating to be the one who comes in after the cycle has started and has to pick up the pieces that were left behind by the promoted manager!

In the following chapters we discuss changes you can make in the work processes. These changes will result in improvements in effectiveness that will allow cost reductions and utilization increases. This approach to cost reduction and reliability allows you to move to pacesetter performance—operation at the level of lowest *sustainable* cost.

Definitions

We begin this book with a set of definitions, since it is important to have an understanding of the terms discussed in the following chapters. The following are the definitions we use when we work with our clients. These may differ from your definitions, but we want to be clear in how we define the work.

Benchmarking data. This data typically is provided by an independent auditing company and is used to measure your performance relative to similar industries (your competitors). In our examples, it references data such as maintenance costs per unit of production capacity, mean time to failures, etc. Since most of our work has been in the refining and chemical industries, we are most familiar with the benchmarking work done by Solomon, Independent Project Analysis, and Townsend.

Turnaround maintenance. This refers to maintenance work that is performed while a manufacturing plant or production line is shut down on a scheduled basis. Frequency of planned shutdowns for plants or production lines varies from a few months to several years. Since production is lost during a turnaround, it is usually the most costly method of doing maintenance. (It has the biggest impact on both cost and production.)

Routine maintenance. This refers to all maintenance completed at a facility that is not included in planned turnarounds. Routine maintenance work usually does not impact production. However, routine maintenance does include unplanned shutdowns (unscheduled turnarounds), and these do impact production.

Routine maintenance includes predictive maintenance, preventive maintenance, and concepts such as total productive maintenance.

Predictive maintenance. This is a subcategory of routine maintenance. Predictive maintenance includes all the work that is completed based on the condition of the equipment; *i.e.*, when monitoring data indicates the need for a repair. This is considered the best technique for maintenance as it allows the maximum time between repairs, prior to a catastrophic failure. *Smart technologies* are now available for most types of equipment to monitor equipment condition and alert you when maintenance is required.

Preventive maintenance. This is a subcategory of routine maintenance. Preventive maintenance is defined as the tasks completed to interrupt a failure mechanism. It includes filter changes, lubrication changes, bolt torque, assessment, etc. Our definition does not include parts replacements due to wear, since that is considered a repair task, not a preventive task.

Total productive maintenance. Appropriate people execute the appropriate work. For example, we would have operations personnel execute as much minor maintenance as they possibly can, since they are already on site 24/7. We would have mechanics do only the work that required their level of expertise. We often ask the question of non-maintenance personnel, "Would you pay someone to come to your home to do the task that you ask maintenance to do here at work?" If they would not pay to have it done at home, they probably should not pay someone else to do it at work.

Work process metrics. These are measurements of employee/manager activities that drive desired business results.

Results metrics. These are measurements of desired business results.

1

BENCHMARKING DATA

I f you are lucky, you are in a business where there is readily-available benchmarking data. It is important to understand that relative to benchmarking data, half the organizations in the study will be above-average performers and half will be below-average performers, as indicated by the metrics benchmarked.

Most of you are in businesses that produce commodities. The top performers (we call them pacesetters) produce these commodities at the lowest sustainable cost. This allows them to remain profitable during the tough times and to make a lot of money during the good times.

Usually the first order of business if you are a below-average performer is to discredit the bench-marking data. Typically the second order of business is to try to explain why your performance is due to things that are outside your control. Past excuses we have heard included

- Tougher environmental rules (California)

- Cold-weather climate (plants located outside the Sunbelt)

- Old equipment (some plants have the philosophy that the equipment will run at its worst the day it enters the plant!)

- High-cost labor market, etc.

Once you are past these first two orders of business, you then face the shock and disbelief that you could be as bad as you are. Next it is time to understand why you are where you are, and what can be done about your situation.

First, you need to understand the gaps in your maintenance and reliability performance. You might find that you are a top performer in routine maintenance but at the bottom of the heap in turnarounds. You could be a top performer in turnarounds and rank the lowest in routine maintenance. Or, you might be at the bottom in both categories. Perhaps you are at the lowest level in reliability but at the best group for maintenance costs (under maintaining). If you are a top performer in all categories, then you probably do not need to read this book!

It also is important to understand key drivers in your business as well as in your country. Some plants could cut manpower to reduce costs via improved work processes, but the country desires full employment. Therefore, manpower reductions from improved efficiencies are not desirable. Make sure you understand your business drivers before you proceed with any improvement efforts. You could be wasting your time.

The benchmarking data in Table 1–1 compares pacesetter data to the routine maintenance performance of a fictional facility. There are three types of plants in this example. Within the types of plants, the sizes vary. The benchmark cost data is normalized based on the design throughputs for each plant. The actual plant performance for 2002 and projected targets for 2003 are compared to the benchmark data in our example. Where there is good performance relative to pacesetter performance, the background is dark gray. Poor performance relative to pacesetter performance has a light gray background. Before

Table 1–1 Sample Benchmarking Data: Routine Maintenance

| Production Plants/ Lines | Unit Size (M#/Hr) (MB/D) | Benchmark Data | | | Plant Data Annual Routine Maintenance (M$) Y2002 Actuals | Plant Data Annual Routine Maintenance (M$) Target Y2003 |
		Avg. Age (Years)	Mechanical Availability	Annual Routine Maintenance (M$)		
Unit A-1	75	24	97%	$ 5,000	$ 9,300	$ 7,000
Unit A-2	90	24	97%	$ 6,000	$ 7,500	$ 6,500
Unit C-1	50	19	94%	$ 9,000	$10,000	$ 8,500
Unit C-2	60	19	94%	$10,800	$15,000	$10,500
Unit C-3	70	19	94%	$12,600	$12,500	$12,000
Unit D-1	45	15	98%	$ 500	$ 900	$ 800
Unit D-2	35	15	98%	$ 389	$ 790	$ 600

arriving at any conclusions, the data should be evaluated to make sure that the differences in performance are statistically significant.

Based on a quick analysis of the data, the plants did not perform at pacesetter levels except in unit C-3 in 2002. Even more disturbing, their plans for 2003 show that they will not reach pacesetter status in any of the A units and D units. It appears they are planning to get their act together on their C units.

This data would indicate there are opportunities with the work processes for routine maintenance in units A and D. We would use our routine maintenance model to understand these units more clearly. We will discuss the routine maintenance model in detail in our next chapter.

The benchmarking data in Table 1–2 compares pacesetter turnaround data to the maintenance performance of a fictional facility. There are three types of plants in this example. Within the types of plants, the sizes vary. The actual plant performance for the most recent turnarounds is tallied in the last two columns. Where there is good performance relative to pacesetter performance, the background

Table 1–2 Sample Benchmarking Data: Turnaround Maintenance (Duration and Interval)

Production Plants/ Lines	Unit Size (M#/Hr) (MB/D)	Avg. Age (Years)	Benchmark Data			Plant Turn-around Interval (Months)	Plant Turn-around Duration (Days)
			Mechanical Availability	Turn-around Interval (Months)	Turn-around Duration (Days)		
Unit A-1	75	24	97%	50	27	31	34
Unit A-2	90	24	97%	50	33	29	28
Unit C-1	50	19	94%	30	33	33	28
Unit C-2	60	19	94%	30	35	33	30
Unit C-3	70	19	94%	30	35	30	25
Unit D-1	45	15	98%	60	23	24	37
Unit D-2	35	15	98%	60	21	24	32

Notes: Turnaround intervals and duration are taken from the turnaround business plan document and shutdown scope documents.

is dark gray. Poor performance relative to pacesetter performance has a light gray background. Before arriving at any conclusions, the data should be evaluated to make sure that the differences in performance are statistically significant.

Based on a quick analysis, the turnarounds for the C units are equal to or better than the pacesetters both on duration of the turnarounds and the intervals between turnarounds. For the other units in this analysis, all have had shorter runs and take longer to execute their turnarounds than the pacesetters. The exception is unit A-2, which had a quick turnaround during its last cycle.

This same analysis should be performed on the plans for the future turnarounds to see if the planning efforts are targeted to attain pacesetter performance.

We would use the best practices in the chapter on turnarounds to address improvements to turnaround duration for unit A-1 and the D units. We would use the information in chapter 4 on reliability to address the interval gaps for all the A and D units.

The benchmarking data in Table 1–3 compares pacesetter turnaround data to the maintenance performance of a fictional facility. There are three types of plants in this example. Within the types of plants, the sizes vary. The actual plant performance for the most recent turnarounds is tallied in the last four columns. The annualized data is calculated by taking the actual or projected costs for the turnarounds and dividing it by the years between turnarounds. Where there is good performance relative to pacesetter performance, the background is dark gray. Poor performance relative to pacesetter performance has a light gray background. Before arriving at any conclusions, the data should be evaluated to make sure that the differences in performance are statistically significant.

Our brief analysis indicates that one of the A units is a pacesetter and one is not. For the C units, none have been considered pacesetters, and only one has a plan to achieve pacesetter status during the next turnaround. The D units are not pace-setters and do not appear to have plans to achieve pacesetter performance.

Table 1-3 Sample Benchmarking Data: Turnaround Maintenance (Cost Data)

Benchmark Data

Production Plants/Lines	Unit Size (MB/D) (M#/Hr)	Avg. Age (Years)	Mechanical Availability	Annualized Turnaround Maintenance (M$)	Last Actual Turnaround Cost (M$)	Annualized Last Actual Turnaround Cost (M$)	Next Turnaround Planned Cost (M$)	Next Turnaround Annualized Cost (M$)
Unit A-1	75	24	97%	$2,000	$ 9,000	$3,484	$ 7,000	$ 2,710
Unit A-2	90	24	97%	$2,400	$ 5,500	$2,276	$ 5,300	$ 2,193
Unit C-1	50	19	94%	$5,000	$17,000	$6,182	$13,000	$ 4,727
Unit C-2	60	19	94%	$6,000	$23,000	$8,364	$21,000	$ 7,636
Unit C-3	70	19	94%	$7,000	$21,000	$8,400	$33,000	$13,200
Unit D-1	45	15	98%	$1,200	$ 4,300	$2,150	$ 3,400	$ 1,700
Unit D-2	35	15	98%	$ 933	$ 2,700	$1,350	$ 3,100	$ 1,550

Notes:

1. Fall-downs or short duration turnarounds are included in the routine maintenance category, not the turnaround costs.

2. Annualized data is calculated by taking actual maintenance costs or projected maintenance costs and dividing it by the number of years between turnarounds (intervals).

3. Data for the intervals comes from the previous table for turnaround durations and intervals.

The information in the chapters on both reliability and turnarounds would be utilized to address the identified gaps from Table 1–3.

The benchmarking data in Table 1–4 compares pacesetter reliability performance to the reliability performance (mechanical availability) of a fictional facility. There are three types of plants in this example. Within the types of plants, the sizes vary. The method of calculating the mechanical availability is explained in Note 1 in Table 1–4. Where there is good performance relative to pacesetter performance, the background is dark gray. Poor performance relative to pacesetter performance has a light gray background. Before arriving at any conclusions, the data should be evaluated to make sure that the differences in performance are statistically significant.

Based on the data, the C units should be at pacesetter mechanical availability for the run, assuming that they do not lose too many days due to falldowns during their normal run cycle. Unit A-1 is already behind pacesetter availability just due to the short interval and long duration of its turnarounds. Unit A-2 has to have a perfect run to stay at pacesetter

Table 1–4 Sample Benchmarking Data: Reliability

Production Plants/ Lines	Unit Size (M#/Hr) (MB/D)	Benchmark Data		Plant Turn-around Interval (Months)	Actual Turn-around Duration (Days)	Calculated Mechanical Availability
		Avg. Age (Years)	Mechanical Availability			
Unit A-1	75	24	97%	31	34	96%
Unit A-2	90	24	97%	29	28	97%
Unit C-1	50	19	94%	33	28	97%
Unit C-2	60	19	94%	33	30	97%
Unit C-3	70	19	94%	30	25	97%
Unit D-1	45	15	98%	24	37	95%
Unit D-2	35	15	98%	24	32	96%

Notes: Best attainable mechanical availability is calculated by dividing turnaround days (duration) by the years between turnarounds (interval). This value is then subtracted from a calendar year to calculate the percentage of days the plant is available (mechanical availability).

performance. The D units are starting off already behind pacesetter performance due to the short interval and long duration of their turnarounds.

Again, the information in chapter 3 on reliability and chapter 4 on turnarounds would be utilized to address the identified gaps from Table 1–4.

Benchmark data providers can provide insight into your performance in general. However, the data providers often have trouble helping you close your gaps relative to pacesetters. In the following chapters, we will help you embark on path to becoming a pacesetter in maintenance and reliability. But first, we will address performance improvements for routine maintenance, one of our favorite topics, in the next chapter.

2

ROUTINE
MAINTENANCE

We define the term *routine maintenance* as all maintenance completed at a facility that is not included in planned turnarounds. Routine maintenance work usually does not impact production. However, routine maintenance includes unplanned shutdowns (unscheduled turnarounds), and these do impact production. Routine maintenance includes predictive maintenance, preventive maintenance, and concepts such as total productive maintenance.

In general, you and your competitors have very similar equipment at your respective facilities, unless you have lost touch with technological advances in your business over the years.

Our experience is that the routine maintenance work you do at your facility will fall into one of the seven categories included in our routine maintenance model shown in Table 2–1. The differences in your performance relative to pacesetter performance probably can be analyzed best by looking at the

Table 2–1 Simplified Routine Maintenance Model

Operator Work	Emergency Repairs	End-of-Life Repairs	Excessive Repairs	Work That Should Not Be Done
		Preventive Maintenance		
		Predictive Maintenance		

proportion of your work that occurs in the categories of our simplified model.

When we work with our clients and try to explain their routine maintenance costs, we find it best to start by looking at our simplified model for routine maintenance. Our analysis consistently finds that the largest gaps between pacesetters and clients are found in the light gray zones of our simplified routine maintenance model. We also find that there are gaps in the dark gray zones when inadequate predictive and preventive maintenance processes are in place, or when the preventive and predictive tasks add no value. We also often find there are inadequate planning and scheduling work processes in place, resulting in ineffective maintenance work.

Our analysis of pacesetters shows that they work to

- Minimize (eliminate) the maintenance work that falls into the categories of maintenance with the light gray background

- Maximize the maintenance work that falls into the categories of maintenance with the dark gray background

- Maximize the effectiveness of the organization to do the maintenance work that falls into the categories of maintenance with the white and dark gray backgrounds (effective maintenance planning and scheduling processes)

In reality, the top performers *just do less work*. Routine maintenance categories as explained in the Table 2–1 model are straightforward. The secret is in implementing the right strategies.

Usually the first effort (from our entire routine maintenance model) that a management team is willing to pursue is to improve the effectiveness of work done in all the categories.

Addressing this first is faulty only because it is not the most efficient place to start. In doing so, we also will be improving the effectiveness of work that *should not* be done by maintenance, as well as improving the effectiveness of the work that they should do.

In our consulting work we emphasize the following key concept :

> *No matter how effective you are at doing maintenance work, if you have maintenance employees doing work that your competitors do not do, you will never catch up. Your competitors will continue to beat you at the game of manufacturing at lowest sustainable cost.*

Why do we use the term *lowest sustainable cost*? Operating at lowest cost would mean sending everyone home and doing no maintenance. But that is not sustainable. Your equipment would stop working fairly quickly. That is why we reference lowest sustainable cost.

We will hit the following point several times in this book—but we do so to get this point across. We recommend that you have your nonmaintenance employees do the work they are capable of doing. This equates to the work they would do themselves at home if a similar problem were to occur there (total productive maintenance).

Total Productive Maintenance

The few brief examples that follow may seem ridiculous, but we are trying to help you open your eyes to the concept of *total productive maintenance.* It is not unusual in the many plants we visit to see an operator write a work request to change a light bulb. Yet this operator would never call an electrician to his home to do that same task. It is also not unusual to see operators requesting maintenance to check oil levels, yet they check and change the oil in their own cars at home. Pacesetters expect their people to do what they are capable of doing in their discretionary time. That is the concept of *total productive maintenance.* This model includes various work categories such as *operator work, emergency repairs, excessive repairs,* and *work that should not be done.*

OPERATOR WORK

Total productive maintenance touches on the issue of operator "down time" (well, I have another term for it, but it's not exactly appropriate for publication!). In the power and petrochemical industries, there can be significant periods of inactivity for an operator—especially during the nighttime hours and weekends. If you have an opportunity to go into a control room of an industrial facility, take a look around and you will see what I mean. Pacesetters attempt to minimize these inherent inactivity periods by implementing total productive maintenance concepts.

EMERGENCY REPAIRS

Industry data indicates that work completed on an emergency basis is the most expensive way to do routine maintenance. It also has the lowest reliability of any repair that is made. An emergency repair will more likely require rework than a job that is well planned and staffed appropriately with the right people and correct repair parts.

EXCESSIVE REPAIRS

Excessive repairs are sometimes due to rework from emergency repairs and to what we refer to as *worst-actor equipment*. We discuss the concept of worst actors in chapter 4, "Reliability." The issue regarding worst actors is the need to eliminate the worst-acting equipment by having resources focused on solving these repetitive problems. In other words, make the work go away! Eliminate the excessive repairs!

WORK THAT SHOULD NOT BE DONE

We also find that to be a pacesetter, you have to stop doing things that do not improve production or eliminate costs. We refer to it as *work that should not be done*. If it is a task that will make life easier for your employees, but will not reduce costs (reduce manpower) or increase production, you probably should not be doing that work. One of my favorite examples was an operator request for a natural gas line to be run to the control room for the new gas barbecue! Pacesetters do the right work effectively: they perform jobs that will increase or maintain their production rates or reduce their costs.

Predictive Maintenance. We find that pacesetters attempt to maximize the percentage of maintenance work they complete using predictive maintenance techniques. Predictive maintenance includes all the work that is completed based on the condition of the equipment—*i.e.* when monitoring data indicates the need for a repair. Predictive maintenance also includes the equipment monitoring tasks. It is considered the best technique for maintenance as it allows the maximum time between repairs, yet precludes catastrophic failures. "Smart" technologies are now available so that most equipment can monitor equipment condition and provide alerts when maintenance is required.

Preventive Maintenance. Pacesetters also attempt to maximize their percentage of maintenance completion using preventive maintenance techniques. Preventive maintenance includes the tasks completed to interrupt a failure mechanism. It would include filter changes, lubrication changes, checking bolt torque, ietc. Our definition does not include parts replacements due to wear since that is considered a repair task (Time-based maintenance), not a preventive task.

We often find that in an attempt to increase the work defined as preventive maintenance, non-value added tasks that do not interrupt any failure cycle are completed under the guise of preventive maintenance. It is important to identify these non-value added tasks and eliminate them as they do not reduce maintenance costs. They only increase costs by increasing the amount of work the organization must do with no benefit to the bottom line.

Routine End-of-Life Repairs. All equipment has a finite life. Routine repairs to equipment (equipment that has given a very acceptable run length) fall into this category. We would consider most repairs that are not predictive or preventive in nature to fall into this category. This category is our "catchall" for repairs to equipment that gave us a reasonable run (not an excessive repair).

EVALUATING PRODUCTIVE MAINTENANCE

We assist our clients to attain top performance in maintenance and reliability and to increase production from their existing facilities. In doing this, we first try to correlate their performance relative to our routine maintenance model. Spending capital to achieve desired results should always be the last option considered.

Our first step is to identify the size of the prize in each of the categories and to obtain management team buy-in on the categories they are willing to address.

Contrary to popular belief, the majority of this analysis can be accomplished by a few knowledgeable people in a period of a week or two. This is true if they are given access to the right data and ability to interview the appropriate personnel and observe work in action in the plant.

In our experience, most management teams are willing to go after improvements in their planning and scheduling processes. But they are often not willing to address the total productive maintenance category (operator work) for their operators. This can be understandable depending on their relationship with these key employees, their operations staffing philosophy (staff for emergency versus staff for sustained operations), and their union relationships, etc. However, it does need to be addressed at some point if pacesetter performance is the ultimate objective.

The successful implementation of a maintenance planning and scheduling process results in immediate benefits of cost reductions and reliability improvements. We have found that after seeing these benefits, the management team becomes more willing to take on the more sensitive categories included in our model.

The issue of unnecessary work is initially taboo with management teams as well. We have found this reaction puzzling and have no explanation for it. But again, the benefits associated with a successful implementation of maintenance planning and scheduling processes help the management team become more willing to address this category, too.

Most often we find that a management team wants the first order of business to be the improvement or creation of a maintenance planning and scheduling process. In these cases we focus on rapid implementation of planning and scheduling improvements. After recommending going after other categories of work in our model first, we do not waste time debating operator work or the other categories in our model as being more important. We get to work to implement the endorsed strategy quickly.

Tackling planning and scheduling first does make implementation a little bit more difficult. This is because you are trying to improve processes for more work than should actually be done by maintenance. But it is better to get started making some improvements versus endlessly debating where to start. (Clarification: the debates are value added to the consultant as they are billable during the debates, but they are not value added to the client!)

However, making the work go away will always generate a faster and greater return versus improving the ability to do work more effectively. This is true especially when your competitors are not doing that work at all.

The majority of our clients have an acceptable maintenance backlog management system in place. What most do not have is an adequate maintenance planning and scheduling module as part of their backlog system. Some of our clients have tried to implement planning and scheduling with consultant assistance, but the implementation went on for more than a year. Thus, they did not capture any of the financial benefits from the effort. They felt they were more effective, but the bottom-line impact was not measurable! The consultant did, however, see benefit, as the time was billable during the entire year.

Our focus is not on selling software, but making do with what the client has in place. We have found that in every case, an Excel workaround could be implemented to convert an acceptable backlog system into an acceptable planning and scheduling system. This system could be developed with appropriate documentation and training with just a one-week effort. The next week implementation would begin. We have also found that it should only take about four weeks per operating area to implement a planning and scheduling strategy. The four weeks time to implement is set to address training for the four operating crews that typically man an operating unit 24/7. It allows adequate practice in the process for the maintenance staff as they go through the same process each of the four weeks.

Our expectation is that a single planner/scheduler should be able to manage the work for a crew of around 20 mechanics. This includes planning the work, procuring materials, and working with their operating counterparts to create a weekly work schedule for a maintenance crew. We would expect the Excel schedule to balance daily the available work hours against available work by craft. We would expect to see schedule loading at levels no less than 100% of available manpower hours each day by craft.

It is not unusual to find that at the end of four weeks of implementation, there is not enough work (*i.e.,* jobs that operations can release, materials available, or jobs planned properly) in the backlog to fill a weekly schedule for the existing maintenance manpower. It is then up to management to address what to do with the excess manpower. We discuss this concern with management prior to starting implementation so they will be ready to deal with these issues. However, it is not unusual that management is surprised when at week three or four we tell them we have a problem, and it is time to deal with excess manpower.

If you are not ready to deal with manpower reductions, we recommend not wasting your time implementing improvements to your maintenance effectiveness. Manpower reductions are where improved effectiveness has impact on bottom-line costs.

3

TURNAROUND MAINTENANCE

W e have defined *turnaround maintenance* as the work that is completed while a manufacturing plant or production line is shut down on a scheduled basis. Frequency of planned shutdowns for plants or production lines varies from a few months to several years. Since production is lost during a turnaround, it is usually the most costly method of doing maintenance. It has the greatest impact on the bottom line of a profit and loss statement.

It is important to understand the best process (relative to profit impact) for a particular plant or facility to do turnarounds. This will vary depending on how you make your money. One company that we worked with had varying approaches to doing turnarounds based on their differing marketing strategies from site to site.

For instance, one facility was in a pure commodity business where all products were sold into the spot market. They found that it was best for their bottom line (profitability) to extend the time between turnarounds as long as possible. One complex ran five years between turnarounds. It was the cheapest strategy for that company to do its turnarounds on an annualized turnaround cost basis.

Another facility found that the largest expense associated with its turnarounds was to buy replacement product on the spot market to keep its contracted customers supplied. This facility's approach was to do turnarounds more frequently, limiting turnaround downtime (turnaround duration) to match available inventory. This strategy eliminated the facility's largest expense. They did not have to purchase product on the spot market to meet contract commitments as long as they were able to complete the turnaround prior to running out of inventory. The annualized turnaround maintenance cost for this approach was more expensive than pacesetter turnaround maintenance. However, the total cost (annualized turnaround maintenance expense plus product replacement) was much less, with less impact to the bottom line, for this particular case.

It is recommended that the impact of each strategy on the performance of your business be studied in order to select the optimum strategy. In most cases, there is an effort to increase the time between turnarounds (the interval), and an effort to reduce the duration of the turnaround. This is all driven by our underlying concept. Simply stated, *since production is lost during a turnaround, it is usually the most costly method (biggest impact to your bottom line) of doing maintenance.* Therefore, there should be an effort to understand why you perform turnarounds so frequently, and why they last as long as they do. Pacesetters go after these limitations to achieve continuous improvement.

For example, one plant was required to shut down every two years due to steam generation permits (they had steam coils in their furnace convection sections). They modified their furnaces by eliminating steam coils and using the waste heat to preheat their furnace air. This modification allowed them to move their turnarounds to five years from two years and achieve pacesetter status. This reduced their annualized turnaround expense by more than 50%, as well as increased their plant availability as a result of fewer turnarounds.

Another plant found they were limited on the number of contractors they could manage and spent considerable effort on reducing the work scope for their turnarounds. They made modifications (such as installing additional valves for isolation and bypass) to allow equipment to be taken off-line during a run. This turnaround work then became part of their routine maintenance (or predictive) program. With the reductions in work scope, the turnarounds were more manageable.

We also note that pacesetters document what went well and what went poorly following a turnaround. This information is critical to the next team that takes on the next series of pacesetter turnarounds.

Some do not take the time to document the lessons learned. Or they do not take the time to study the lessons documented by their predecessors. These people are doomed to repeat their mistakes. Pacesetters learn from their mistakes and do not repeat them—a pretty simple concept.

Benchmarking data should be used to assist in setting goals or targets for cost, interval, and duration. Often benchmark data is used after the turnaround to see how well a turnaround was implemented. This occurs more frequently than using the same benchmark data proactively to assist in setting goals and objectives in the early stages of planning the turnaround.

Using benchmarking data to set objectives rather than to judge performance after the work is done is another strategy to achieve pacesetter performance.

Also important is having the core turnaround team in place to address the multiple phases for turnaround planning and execution at the appropriate times. We find it is all too common that the facility management will wait too long to staff the core turnaround team. This delay will almost always guarantee unsatisfactory results for the turnaround. The key deliverables in the various phases of the turnaround planning process are delayed, or worse yet, not completed. Pacesetters begin their turnaround process 18–24 months prior to the start of the turnaround execution.

It is also important to bring the turnaround execution line supervisors on board early. Since they will execute the turnaround work, they require adequate time to review the shutdown schedule, their work packages and areas of responsibility, and the order in which the work will be completed. Again, we find all too often that a facility assigns the turnaround line maintenance supervisors to their posts the day before the turnaround execution begins. Excellent planning efforts can easily be unraveled by less-than-adequate execution.

Maintenance organizations commonly focus their critical-path scheduling efforts on minimizing the length of time between when the operators turn the plant over to maintenance and when maintenance gives the plant back to the operators to begin their start-up process.

Pacesetters focus their critical-path scheduling efforts on minimizing the total time between the feed-out to feed-in process. This includes operator shutdown procedures, plant cleanup, plant handoff (both ways), operator start-up procedures, and the actual maintenance activities. Many operator activities can overlap the maintenance activities if the critical path schedule looks at the entire feed-out to feed-in processes. Focusing on this area can provide significant potential for reducing total facility downtime.

It is important for a facility to have a turnaround philosophy in place to cover how the facility will conduct its turnarounds (including the plan, schedule, execution, and capturing any lessons learned). It should address the use of benchmark data and objectives include the following:

- Strive to attain longer runs for processing units (longer intervals between turnarounds).

- Do not open equipment on a turnaround to "just take a look." Have valid reasons to open equipment as this will only extend turn-around duration and add to the manpower resources required for the turnaround.

- Analyze all failures and repair them appropriately to prevent recurrence.

- Differentiate between types of turnarounds with appropriate philosophy statements, such as

  - **Planned turnarounds.** Use best practices to address planned turnarounds. Focus on work that can be done only when the unit is off-line.

  - **Planned catalyst changes.** Use best practices for these small turnarounds to address the blinding and catalyst change. The turnaround effort will focus solely on the catalyst change to quickly return the plant to operation with no maintenance performed during turnaround.

  - **Unplanned shutdowns.** Unplanned shutdowns are the most disruptive to business, resulting in costly unplanned maintenance. Due to the impact on production and costs, the focus should be on completing a quality repair to the broken equipment that caused the shutdown and to return the unit to service. Attempting to do more than

repair the items that caused the shutdown will be very expensive (since the
shutdown work has not been planned per
best practices). It will usually add to the
downtime.

–   **Feed outage due to business reasons,
    not due to equipment problems.** The
    unit will be idled, or parked, until there is
    a business reason to return it to opera-
    tion. No turnaround maintenance work
    will be completed on a parked plant.

We have attempted to document on the
following pages many of the key steps necessary to
guarantee a pacesetter-quality turnaround.
However, if you will not attain pacesetter status
before the turnaround work, you should not be
surprised if you are not at a pacesetter performance
level afterwards.

Pacesetters typically have multiple gates or
phases in their turnaround planning, scheduling,
and implementation processes. They have identified
the associated deliverables that assist in attaining
pacesetter performance and have assigned them to

the appropriate gates or phases of the turnaround process. It is very important that the deliverables from one gate or phase of the turnaround planning and execution process be in place and endorsed by management prior to moving to a subsequent gate. It is often difficult to catch up if you let things slide in the early planning periods.

We have found that the first few times a new turnaround process is followed it is best to have external resources on a review board. These external resources can verify readiness to move from one gate or phase of the turnaround process to another. They will more often use what we call *cold eyes* to review the quality of the deliverables for each gate or phase before endorsing the move to a subsequent gate or phase. It is too easy for an internal management team to say, "Let's keep going—we will catch up later" when the turnaround team falls behind the scheduled events of the turnaround planning process schedule. The catch-up rarely happens.

In the next section is a summary of deliverables we find necessary to guarantee pacesetter turnaround performance. The items listed are self-explanatory. These are the basics that we already know we should be doing. We just need to make sure we find the time and discipline to make the right things happen at the right times. Top performers may disagree on the exact timing for the gate or phase for each of the deliverables. However, they do agree that they have to occur prior to moving to the next gate or phase identified in their specific turnaround process.

You should find few surprises in your turnaround performance relative to your predicted cost and duration if you are honest with yourself during the planning and estimating stages of this process. This also depends on completing your of deliverables at the appropriate milestones.

Do not expect pacesetter performance if plans do not indicate the appropriate pacesetter levels for turnaround intervals, duration, and expenditures during the early periods of the planning process.

Deliverables in the Turnaround Process

The following is a partial list of deliverables in the turnaround process.

BUSINESS PLANNING DELIVERABLES

- Long-range turnaround forecast based on marketing philosophy

- Maintenance turnaround targets versus pacesetter performance (plant mechanical availability, turnaround cost, annualized turnaround cost, and turnaround duration)

- Gap closure plans to attain pacesetter performance

TURNAROUND PLANNING AND SCHEDULING DELIVERABLES

- Timeline with milestones and deliverables for the turnaround planning effort

- Audit of lessons learned and best practices from past turnarounds

- Audit of OSHA compliance requirements (process hazards analysis, prestart-up safety review, etc.)

- Audit of historical plant turnaround documents

- Turnaround philosophy (target for next run: production rates, mechanical availability, interval, product quality, etc.)

- Critical-path schedule including operations shutdown, cleanup, and start-up requirements versus pacesetter targets

- Estimated turnaround duration (feed out to feed in) versus pacesetter targets

- Turnaround work list

    - Plant upgrades

    - Environmental requirements

    - OSHA-required work

    - Reliability improvements

- – Legal compliance requirements

- – Operator, engineer, and mechanic suggested changes/improvements

- Turnaround cost estimate versus pacesetter target

- Manpower forecast and supplemental contracting plan for turnaround (Is it manageable?)

- Resolution of constructability issues

- Work orders issued for all work

- Long delivery materials tracked for on-time arrival

- Environmental, safety, fire, and health turnaround documents published

- Capital and expense expenditure forms for the planned work, identifying steps in place to extend turnaround intervals

PRETURNAROUND WORK DELIVERABLES

- Turnaround execution team mobilized for work package (detailed) review to understand implementation steps

- Cost tracking and reporting in progress

- Complete prefabrication work where possible

- Stage/prepare temporary connections for cleanup/ blinding

- Stage/prepare required tooling

- Conduct mechanic and operator training for changes or upgrades to the plant/production line

- Conduct any required contractor safety training

TURNAROUND EXECUTION DELIVERABLES

- Unit shut down safely and cleaned per schedule—coordinated with critical-path maintenance work

- Supervisors in the field at job sites at least 90% of the time

- Daily coordination meetings

    - Schedule reviews and updates

    - Cost reviews, updates, and reports

    - Additional work review and proper authorizations as required

- Prestart-up safety review process ongoing as turnaround work progresses

POST START-UP DELIVERABLES

- Documentation of lessons learned (positive and negative)

- Documentation of recommendations for future turnarounds

4

RELIABILITY

Many of the plants that we work with feel that facility reliability is everyone's business. And it is. However, the problem with everyone owning a process is that nobody owns the process.

We work with many plants, and their reliability varies significantly. For example, one facility has pump mechanical seals with an average mean time to repair of 10 years! Another facility with the same types of equipment has a mean time to repair of 6 months for their mechanical seals. That means that the second plant does 20 times the maintenance on the same types of equipment...20 times as much work! And, the plant with the 6-month mean time to repair does not believe the data! No one else could be getting 10 years on seals when they are only getting a half year! We find these large variations in mean time to repair from site to site for a large variety of equipment. This includes machinery, electrical equipment, instrumentation, fixed equipment (columns, drums, vessels, heat exchangers, reactors, etc.), and analyzers!

We also find many plants with reliability organizations spending their days *fighting fires*. They are constantly called to assist with a repair on a piece of equipment that could be handled adequately by a maintenance mechanic without the reliability organization's involvement. These facilities usually have substandard reliability.

We find that the top performers have reliability organizations that work with their operators to teach them how to operate their equipment with excellence. This often requires additional training for operators. However, is well worth the effort if you can get them to use their new training to prevent failures by properly caring for the equipment.

Top performers work with their engineering organizations (technical organizations) so that new equipment is purchased based on total cost of ownership. They consider more than just the initial purchase, which usually is made on a low-bid basis. Lowest installed cost often means lowest quality and a higher cost of maintenance.

My dad used to say, "You get what you pay for." He was right! A 10-year mechanical seal may cost a couple of thousand dollars more than a 6-month mechanical seal. However, a total cost-of-ownership analysis would quickly indicate that you should spend the additional money up front to get the correct seal!

There could be another problem in the organization as well. The engineer specifying the equipment (seals, for instance) may not know there is a better alternative out there for the application. If this is your situation, you need to consider establishing an alliance with top manufacturing companies for your equipment. They will tell you there are differences in their products—what they bid to make a sale versus their top performing products for each application. And, this is the case for all types of equipment!

Top performers work with their maintenance mechanics to improve the techniques used to make repairs properly. These repairs are in line with certain specifications and checklists to guarantee a long run time before a future repair is required.

They also have a dedicated reliability organization that is required to follow key fundamentals. They are measured (*metrics*) on how well they perform work processes to improve reliability as well as results (mean time to failure indices).

Key Concepts to Reliability

There are six key concepts within our reliability model. Many revolve around the concept of worst actor equipment. The 80/20 rule applies to reliability as well, and it is probably closer to a 90/10 rule. In other words, 10% of your equipment is consuming 90% of your maintenance resources and causing 90% of your lost profits due to plant down time or production line rate reductions. The key reliability concepts are record keeping, design, review, equipment surveillance, quality assurance, protection systems, and auditing.

RECORD KEEPING

The first key reliability concept is record keeping. You must have good records to be able to solve long-term repetitive problems (your worst actor equipment). However, do not spend a couple of years gathering data (record keeping) before you tackle a few of your worst actors.

WORST ACTOR DESIGN REVIEW

The second key reliability concept is to complete a design review of your worst actor equipment. We have found that the best people to address worst actors are mechanics who have been trained to solve problems. Most engineers do not stay in a position long enough to become proficient in solving difficult equipment problems. They are usually thinking about their next promotion on the way to becoming the next CEO.

We have found that the resources necessary to solve these difficult recurring problems must be dedicated resources. The mechanics need to work full time solving these worst acting pieces of equipment, and it helps if they are in the same job for several years.

There are many techniques to attain high reliability and resolve worst actor equipment for a plant/production line. One methodology is called reliability-centered maintenance (RCM), and a variety of books and consulting services are available for information on the topic. The approach is to review each part from a piece of equipment and to analyze the impact of its failure. If there is no impact, RCM strategies recommend letting it fail. The approach also looks at failure detection techniques for each part in a piece of equipment and at sparing philosophies if the failure is critical and cannot be predicted. The technique is quite time-consuming and can be manpower intensive. We typically recommend this technique on specific pieces of equipment that have significant potential impact on production or costs due to failures.

A technique often overlooked for identifying the root causes of a failure is a tool called multiple correlation analysis (MCA). There are several providers of this statistical tool. We have found HITS-MCA developed by Harold S. Haller & Company to be the tool of choice for our multivariate analytical work. Dr. Haller provides training in the use of this methodology, requiring class attendees to bring along a problem that will be solved as part of the class. It is one of the better classes available, as you solve a recurring problem at the same time you are obtaining training on the use of a sophisticated problem resolution tool.

We have used HITS-MCA at several clients' facilities to significantly reduce maintenance work by eliminating recurring problems. In one example, HITS-MCA identified a sweet spot for operating a plant. The plant had been plagued with three or four one-week shutdowns per year to clean a heat exchanger. A sweet spot was identified for operation of the disengager vessel that completely eliminated the heat exchanger plugging problems. The last we checked, the plant had been running continuously for two years with no shutdowns.

In another example, HITS-MCA identified a cold wash system as the culprit for short runs on a centrifuge. When the centrifuge was down, the plant throughput was severely limited. By correcting the wash temperature, the cause of the short runs was eliminated.

A third example concerned a problem in a chemical reaction. HITS-MCA found a variable in the plant that was linked to making off-test product. When the variable was investigated, it was found that an analyzer that generated the data for the variable was not set up properly in the computer. Consequently it gave operators corrupt information on which to base their corrective actions to control the plant. Again, once the correction identified by HITS-MCA was made, the off-test product problem was eliminated.

In all three examples, HITS-MCA was used to identify key variables from a list of a dozen or more variables that produced the worst actors. This equipment was not only causing additional maintenance work, but also had a significant impact on production.

Regardless of the tools you choose, we find that mechanics who have been applying their trade to repair your equipment problems over the years are a valuable resource. Usually they already have several ideas that will eliminate many of your recurring problems. Moreover, these craftsmen could be the best candidates to fill any vacant reliability analyst positions in your organization.

Why didn't these same craftsmen solve the problems while they were applying the skills of their trade? It usually comes down to a time issue or a perceived pressure from management to hurry the repair. The problem may be a part that is not readily available during a repair. Consequently, each time they make a quick fix and quickly reassemble the equipment because there is pressure to get the unit back online. Or, it may be a problem that the individual just does not have the time to resolve while working in the role of a mechanic. Whatever the reason, the problem does not get resolved until you put individuals into dedicated problem-solving positions and give them the time and resources to make the necessary improvements.

We usually expect to see a reliability analyst resolve 1 worst actor each month, working on 10 or so at any time. We have found it necessary that the number of worst actors solved each quarter be one of the key performance indicators for a reliability organization. This emphasis via metrics demonstrates to the organization that it is one of their key responsibilities to resolve these costly problems (and make the work go away).

EQUIPMENT SURVEILLANCE

The third key reliability concept is equipment surveillance. A key metric is how many failures were unexpected. If a trained surveillance technician is allowed to do the surveillance routes as scheduled, there should be no surprise failures. If there are surprises, you need to question the surveillance technician's level of training, the time available to conduct surveillance, or the target frequency. *Smart equipment* (such as smart transmitters or smart valve positioners) can provide continuous monitoring (surveillance).

QUALITY ASSURANCE

The fourth key reliability concept is quality assurance. For most repairs, a mechanic should be able to provide his own quality assurance (quality control checks using checklists for specific repairs). However, when a repair or improvement to a worst actor is implemented, the reliability representative should be present. The representative's role is to verify that the repeat failure is not due to a repair step in the process and also make sure that any corrections implemented to eliminate a repeat problem are completed properly.

PROTECTION SYSTEMS FOR CRITICAL EQUIPMENT

The fifth key reliability concept is to provide appropriate protection systems for critical equipment. You cannot afford to allow certain pieces of equipment to fail catastrophically. For those pieces of equipment, you should verify that their protection devices are maintained properly and armed. Types of protection include overpressure protection devices (relief valves), electrical relays, vibration detection devices, thrust probes, automatic shutdown systems, and turbine overspeed trips.

AUDITING

The sixth key reliability concept is not popular: auditing. If you want reliability, everyone has a part to play in the process. Someone needs to audit the system to make sure everyone in the organization completes their assigned roles and responsibilities to guarantee the reliability of the equipment.

Someone has to conduct these audits. Someone has to inspect the equipment and monitor the procedures to make sure appropriate equipment repairs and services are completed as required. This responsibility appears to fit best within a reliability organization.

One interesting dilemma that we often see at facilities is confusion on where to begin. There is a lot to do. How do you know where to start? We frequently see ongoing debates on how they will define a worst actor. Some of these discussions have been going on for a couple of years when we arrive at a site. And during these debates, they have not started to address a single worst actor since they cannot agree on a definition.

We also see clients that are 6–12 months into implementing a reliability computer program and are gathering data to identify their worst actors. They are usually a couple of years away from starting to address their worst actor equipment while they wait for the data to accumulate for their analysis!

Whatever computer program you choose, we suggest that a parallel effort be established to select your worst actor equipment. We would recommend that you ask your operators and mechanics for a list of the worst actor equipment—the things that bother them the most. Their list will probably contain 80–90% of the worst actor equipment. Then, get on with solving some of the problems while you begin to gather more accurate data with your new computer system. You will be a couple of years ahead and impact your plant reliability now if you take our advice and follow Nike's slogan—"Just do it!"

We do not want to create a perception that we are against using computer programs for reliability. In fact, we strongly endorse the use of computer programs to track your data and to do your analysis. There are some very good ones on the market. We have our favorite, but again, we do not try to sell software. We just want you to start eliminating some of your worst actor equipment now. Eliminate the problems that generate excessive maintenance work and reduce plant production.

5

METRICS

Many expert consultants are capable of helping clients set up an appropriate system of metrics to monitor performance. When we work with our clients, we insist that we set up metrics for each effort we implement as part of our implementation process. It is our intent that we start gathering data on day one of our implementation effort.

We feel that it is very important to have a mix of work process metrics as well as results metrics. As we previously stated, we define *work process metrics* as the measurement of employee activities that drive desired business results. We define *results metrics* as the measurement of desired business results.

We will discuss the concept of metrics in this chapter and provide examples that could be used in the manufacturing arena to track the concepts we have discussed throughout this book. The metrics we provide are not all-inclusive. In some cases, they may not be necessary. They are offered as a starting place to stimulate thought on the topic.

Routine Maintenance Metrics

When we implement a routine maintenance effort, we first identify what we want to accomplish. In most cases, we want to reduce the perceived size of the backlog of work as well as reduce maintenance costs. In addition, to reduce costs, we want to reduce the number of contractors *coming in the gate*. Management usually (and rightfully so) has a hands-off approach toward company manning levels. This attitude is helpful in getting company resources to buy into the implementation efforts.

Setting up the metrics for results therefore would include a daily count of the contractors in a facility. This is not too difficult—just have the contractor gate guard provide the number each day.

Tracking the backlog size is not too difficult, either. We recommend tracking the number of jobs in the backlog versus tracking the number of hours, as we are looking for trends versus exact numbers here. Estimates for job hours are often inaccurate.

Tracking the maintenance expenditures is usually already a metric at most plants. The only problem is that the financial data is often a month old at best—not too effective for managing the business. Most refer to this as "driving a car by looking in the rearview mirror."

Once your results metrics are established and endorsed, we identify the appropriate work processes necessary to drive the desired results. The work process metrics will be your bellwether indicators for your desired results. We refer to the simplified routine maintenance model as we set up the work process metrics for routine maintenance.

Simplified Routine Maintenance Model

This section discusses the work process metrics we would recommend if you were attempting to improve in all seven categories of work in the simplified routine maintenance model.

OPERATOR WORK

Referring back to our simplified routine maintenance model (Table 2–1 in chapter 2), the top left category is *Operator Work*. Suppose we are trying to reduce the number of *Operator Work* jobs completed by maintenance and have the operators do that work. Then each week we would review the new jobs entered into the maintenance backlog.

We subjectively count the number of jobs that fall under the definition of *Operator Work* for that week and enter the value into a trend chart. This data would then be fed back to the appropriate operations management team for their action.

For each work process metric, we would look for four things:

1. We would expect the key people to understand that particular work process and why we are doing it. We would want them to know what result we are trying to drive with the work process. With maintenance doing less of the sort of work that an operator is capable of doing, we would expect the result metrics (maintenance costs and contractor counts) to decrease. In other words, we would want a clear understanding of the direction the facility is trying to move regarding this work process and corresponding metric.

2. We would verify that these key people (the operators) who have the ability to impact this metric are capable of doing the work that we are requesting of them. (They are competent for the task.)

3. We would verify that there is adequate operator inactivity in the schedules of these key people to do what we desire. If they are

already overloaded with other tasks, it would not make sense to be loading more onto their full plates.

4. Finally, we would make sure that there is a positive consequence for moving the work processes in the proper direction. (Operators are doing some of the work they had pushed to maintenance in the past.)

What are positive consequences? There are many books written on this topic. The problem is that a positive consequence for one person can be a negative consequence to another person. You have to know your people.

Usually, acknowledging a person's behavior with a pinpointed comment regarding their contribution to move a work process in the right direction is a positive consequence. If this acknowledgment is done in front of a group, the potential ribbing from peers may result in more of a negative consequence than a positive consequence for the recognition. Again, you have to know your people.

For each gap closure effort, we would verify that the preceding four points are covered. We would want people to understand what we are trying to accomplish with that particular effort. We would verify that the individuals are capable of doing what we need to have done and that there are adequate resources to do the task (the time to do it as well as the tools and funding to do it). And, we would verify that metrics and appropriate consequences are in place for closing that particular gap.

EMERGENCY REPAIRS

The second category in our simplified routine maintenance model is *Emergency Repairs*. Often the classification of emergency work is abused, as it generates a quick response regardless of the true nature of the work. We would recommend counting the number of emergency jobs requested each week and trending the values. The data then should be fed back to the appropriate operations management team to deal with any abuse of this category of work. A brief analysis of the emergency jobs in the backlog would quickly identify abuse.

EXCESSIVE REPAIRS

Our third category is *Excessive Repairs*. For this metric, we often count the number of worst actors that the reliability team resolves each month and trend these values. We covered the concept of worst actors and excessive repairs in much more detail in chapter 4, "Reliability."

Again, the issue we discuss regarding worst actors is the need to eliminate the worst-acting equipment by focusing resources on solving these repetitive problems. In other words, make the work go away! We feel that a mature program will generate worst actor resolutions at the rate of about one per month per reliability analyst, assuming their other duties allow adequate time to resolve the problems. A number of other reliability activities are discussed in detail in chapter 4.

WORK THAT SHOULD NOT BE DONE

In our fourth category, we again subjectively count the number of jobs requested each week that do not maintain or improve production, or that do not eliminate costs, and we trend those values. If a task will make life easier for your employees, but will not reduce costs or increase production, you probably should not be doing that work. We feed this data back to the appropriate management team to deal with the abuse of this category of work.

PREVENTIVE MAINTENANCE

Here, we trend the percent of maintenance work that falls into the preventive category, and the percent of compliance with the preventive maintenance program.

PREDICTIVE MAINTENANCE

In *Predictive Maintenance* we would trend the percent of surveillance schedules completed on time, and the percent of failures that were predicted by the predictive maintenance effort. Quality surveillance should result in very few surprise failures.

ROUTINE END-OF-LIFE REPAIRS

The final category in our simplified routine maintenance model is *Routine End-of-Life Repairs*. We would be interested in how well the maintenance process was working for this category of work. Metrics would include work schedule loading, schedule compliance, and schedule disruptions. The metrics would actually monitor the effectiveness of the entire routine maintenance work process and not just the routine end-of-life repairs.

In our last two steps, we would ask the following question of each metric: "Is the work process we are measuring truly going to impact our desired result?" If not, we would drop the work process and the metric. (This is an issue mentioned previously: Is the work process really necessary for our desired results?)

We also would ask the question, "If we do these work processes, will they guarantee our desired results?" If the answer is *yes*, we are done with our work process metric identification work. If the answer is *no*, then we have more work to do to identify the remaining work processes and metrics we must capture to guarantee the results we desire. (This is an issue mentioned previously: Are the work processes sufficient to guarantee our desired results?)

Tables 5–1 through 5–12 demonstrate these various processes.

Table 5–1 Sample Routine Maintenance Metrics

Target Improvement Area	Metric	Data Capture Technique	Metric Type
Operator Work	Number of *Operator Work* type work orders entered each week	Count the number of *Operator Work* type work orders entered each week in the maintenance backlog	Work Process
Emergency Repairs	Number of emergency work orders entered each week	Sort weekly backlog by priority and count the number of emergency jobs	Work Process
Excessive Repairs	Number of worst actor resolutions each month	Count the number of worst actor resolutions implemented each month	Work Process

Table 5–2 Sample Routine Maintenance Metrics

Target Improvement Area	Metric	Data Capture Technique	Metric Type
Work That Should Not Be Done	Number of jobs that do not maintain or improve production, or that do not eliminate costs each week	Subjectively count the number of jobs entered into the backlog that do not maintain or improve production, or that do not eliminate costs each week	Work Process
Preventive Maintenance (PM)	Percent compliance with PM schedule	Divide the completed PM tasks by the number of scheduled PM tasks	Work Process
	Percent PM work	Divide the PM hours completed by the total routine maintenance hours consumed each week	Work Process

Table 5–3 Sample Routine Maintenance Metrics

Target Improvement Area	Metric	Data Capture Technique	Metric Type
Predictive Maintenance (PdM)	Percent compliance with the surveillance routes	Calculate the percentage of the surveillance routes that were completed each week	Work Process
	Percent of failures that were predicted	Divide the number of predicted failures by the number of failures each week by equipment type	Work Process
Routine End-of-Life Repairs	Work schedule loading	Ratio of scheduled hours to mechanic available hours	Work Process
	Schedule compliance	Percent of scheduled work completed each week	Work Process
	Schedule breaks	Number of break-in jobs	Work Process

Table 5-4 Sample Routine Maintenance Metrics

Target Improvement Area	Metric	Data Capture Technique	Metric Type
Contractor Counts	Number of contractors in the plant to do routine maintenance	Count the number of contractors in the plant each day	Results
Routine Maintenance Expense	Routine Maintenance Expenditures	Compare routine maintenance expenditures each month versus budget	Results

Table 5-5 Sample Turnaround Maintenance Metrics

Target Improvement Area	Metric	Data Capture Technique	Metric Type
Turnaround Planning Effectiveness	Turnaround team planning phase schedule compliance	Record the percent of turnaround planning process milestones completed on schedule	Work Process
Turnaround Execution Preparation Effectiveness	Days ahead of the feed outage that line maintenance was staffed per plan	Document the days ahead of the feed outage that line maintenance was staffed per plan	Work Process
Effectiveness of Pre-Fabrication Work Efforts	Percentage of pre-fabrication work completed prior to feed outage	Calculate the amount of pre-fabrication work that was completed post feed outage	Work Process

Table 5–6 Sample Turnaround Maintenance Metrics

Target Improvement Area	Metric	Data Capture Technique	Metric Type
Turnaround Schedule Compliance	Schedule compliance	Percent of scheduled work completed versus turnaround schedule each day during the turnaround	Work Process
Contractor Counts	Number of contractors in the plant to do turnaround maintenance	Count the number of contractors on each turnaround every day versus the planned manning levels	Results
Turnaround Cost Tracking	Dollars expended versus target expenditures	Ratio percentage of work completed versus percentage of budgeted dollars expended on the turnaround each day during the turnaround	Work Process

Table 5–7 Sample Turnaround Maintenance Metrics

Target Improvement Area	Metric	Data Capture Technique	Metric Type
Annualized Turnaround Cost	Annualized turn-around cost by plant/unit/line versus pacesetter benchmark	Document projected and finalized costs at the end of each turnaround. Annualize the data based on the next turn-around interval versus pacesetter performance	Results

Table 5–8 Sample Reliability Metrics

Target Improvement Area	Metric	Data Capture Technique	Metric Type
Record Keeping	Percent repairs documented	Count the number of completed repair checklists and divide by the number of repairs completed by equipment type monthly	Work Process
Design Review (Worst Actor Resolution)	Number of worst actor equipments eliminated from worst actor list	Count the number of worst actors reselved via reliability interventions monthly	Work Process

Table 5–9 Sample Reliability Metrics

Target Improvement Area	Metric	Data Capture Technique	Metric Type
Surveillance Effectiveness	Percent compliance with the surveillance routes by equipment type	Calculate the percentage of the surveillance routes that were completed each week	Work Process
	Percentage of failures that were predicted by equipment type	Divide the number of predicted failures by the number of failures each week by equipment type	Work Process
Surveillance (Inspection) Effectiveness	Percentage of plants/lines with inspection plans and strategies	Divide the number of plants/lines with inspection strategies by the number of plants monthly	Work Process
	Compliance with inspection strategies	Track schedule compliance of planned inspections by equipment type monthly	Work Process

Table 5–10 Sample Reliability Metrics

Target Improvement Area	Metric	Data Capture Technique	Metric Type
Quality Assurance (QA)	Repeat repairs (failures in less than 24-month period)	Count the number of repeat failures by equipment type monthly	Result
	Worst actors repairs with appropriate level of QA checks during the repair	Calculate the percent of worst actor repairs with added QA checks	Work Process
Protection Systems	Percent safety systems armed in the shutdown mode	Divide the number of safety systems properly armed by the number of installed safety systems	Work Process

Table 5–11 Sample Reliability Metrics

Target Improvement Area	Metric	Data Capture Technique	Metric Type
Audit Compliance	Percent compliance with reliability strategies in all departments	Calculate the percent compliant with reliability tasks by department monthly (control loops in normal mode, alarms bypassed, alarms tested, lubrication checks, safety shutdown systems armed, spare run program, incidents properly investigated, etc.)	Work Process

Table 5–12 Sample Reliability Metrics

Target Improvement Area	Metric	Data Capture Technique	Metric Type
Plant Reliability	Mean time to repair by various equipment categories	Divide the number of pieces of equipment by the annualized number of repairs each month	Results
Plant Reliability–Lost Profit Opportunities	Dollars lost due to equipment breakdown by equipment type	Total the lost production value by equipment type each month	Results

There are many more metrics to consider when setting up an improvement process. The metrics can become very detailed and cumbersome, with some not worth the effort required to capture the data. These tables were presented as a sampling to generate thought on the reader's part—to stimulate the thinking process to define what metrics will be right for each particular situation. The key to setting the right metrics is in the two statements documented earlier and repeated here due to their importance:

1. *Is the work process we are measuring truly going to impact our desired result?* If not, drop the work process and the metric.

2. *If we do these things, will it guarantee our desired results?* If the answer is *yes*, you are done with developing your work processes and metrics. If the answer is *no*, then you have more work to do to identify the remaining work processes and metrics needed to guarantee the desired results.

6

FRONTLINE
SUPERVISION

No matter how good your maintenance planning and scheduling efforts are for turnaround and other work, it is up to the frontline supervisors to implement the plan with excellence. An excellent plan can be implemented poorly, generating poor results. A poor plan can be salvaged with excellent implementation efforts.

What is required for excellent implementation? The supervisors must have adequate discretionary time to spend with the workforce where the work is being done. I had two types of supervisors working for me:

- Those who had little time to spend with their workforce because of the paperwork associated with the job

- Those who had little time to spend on the paperwork associated with the job because of the time they spent at the job site with their workforce

The latter of the two groups had more successes and delivered excellent implementation efforts. The former group used the paperwork and meetings as excuses to not go to the job sites.

Why would supervisors hide behind paperwork and meetings? Why would management have so many meetings and paperwork for these key supervisors when successful implementation depends on their access to their workers?

My view on the situation is that we have gone a little too far in our efforts to get everyone engaged in the business via our nonstop meetings. Part of the business requires a focus on the business of planning, and another part of the business requires a focus on implementing others' plans. Transfer of planning information to the implementation team does not require continuous meetings.

To understand why some supervisors do not spend adequate time with their workers, it is best to look at the motivational impacts driving their behaviors. What are the consequences for staying in the office and not being at the job site? The weather is usually pretty nice in the office. It's warm if you are in a cold climate, cool if you are in a hot climate, and dry when it is raining. These are all positive consequences for staying away from the job site. What about the consequences at the job site? They can be very negative if you are not up to the challenge of dealing with confrontation. Having to deal with poor performance can be a negative consequence for many supervisors. In short, the natural consequences driving behavior will keep the supervisor away from the job site. These can often lead to poor implementation regardless of the quality of the planning effort.

If you think about a top-performing sports team, their coaches are highly visible during the game. They are not off in a meeting, unavailable during key plays. They are not sitting in an air-conditioned office somewhere while the team plays the game in 90° heat and high humidity. They are *in* the game. Our frontline supervisors are the coaches for the game of maintenance. We need them out there with the players...in this case, the mechanics getting the work done. We need them in the game.

A key role for senior management is to provide adequate consequences for supervisors to be at the job site delivering excellent implementation efforts of excellent plans. You have to know what will be considered a positive consequence for each particular supervisor in order to obtain a desired repeat behavior. In this case, it's supervisory time at the job site.

This time at the job site allows for many important implementation checks:

- Audit the worksite safety and safe actions by the workers.

- Provide positive reinforcement to the workers for a job well done.

- Conduct quality control checks and housekeeping checks.

- Observe the work to look ahead and to provide an early warning of pending problems (incorrect tools at the job site, material shortfalls, etc.). The supervisor can correct these problems to prevent lost time on the job.

You will find that the level and quality of implementation will vary from supervisor to supervisor. Our experience is that the supervisors who spend the majority of their time at the jobsite with their workers deliver the best results.

How much time should they spend with their workers at the job site? A better way to approach the issue is to determine how much time is required for them to be in the office, and how this can be minimized. I would expect them to

- have a 5-minute safety talk at the start of the day with their crew

- spend a maximum of 10 minutes a day on timesheets (if it is not an automated system requiring none of their time)

- spend 20 minutes a day with planners to consider the following day's work and to feedback the day's performance to the planners for planning improvements and job closeout

There may be a few additional tasks that are specific to a particular business. However, we believe it is reasonable for a supervisor to require less than an hour a day on average in the office, away from the job site. For a crew working 10-hour days, this equates to 90% of the supervisor's day being discretionary time. Ideally, the supervisors spend most of their time at the job site with the workers, providing timely feedback on their employees' performance. The supervisors also should be looking ahead for roadblocks that would prevent them from delivering a fair day's work for a fair day's pay—just like a coach on a major league team!

We need to clarify expectations regarding the workforce. What we expect to get from the workforce is a fair day's work for a fair day's pay. We do not expect folks to be "killing themselves" at the job site...but we do expect a fair day's work.

What is a fair day's work? It is what you would expect from a contractor who was doing a job for you at your home and who is receiving an hourly rate from you. We do not think you would expect one to spend the first hour drinking coffee and talking about last night's game if it is your money on the clock. You would not expect them to take several half-hour breaks in the morning and afternoon, or quit an hour before you stopped paying them because it is "too late to start another task." It should not be unreasonable to expect the same level of performance from the company workforce when it is on the company's clock.

We have visited a few companies (call one company "ZCOMP" for this example) where the feeling was, "If you are not working at ZCOMP, you are working too hard!" Pretty interesting message for the management to think about!

It is not unreasonable for a management team to expect a fair day's work for a fair day's pay. Consider the chicken ranch philosophy: Chickens are there to lay eggs. They are called *layers*. When the chickens quit laying eggs, they become *fryers*. As managers, we should be looking for a few eggs each day from each worker. Otherwise, we need to convert them to *fryers* and remove them from the payroll.

With that said, we wish you good luck and happy implementing. Call if you need help.

Bibliography

Daniels, Aubrey. *Bringing Out the Best in People.* McGraw-Hill, 1999.

Moubray, John. *Reliability-Centered Maintenance.* Industrial Press, Inc., 2001

Nelson, Bob. *1001 Ways to Reward Employees.* New York: Workman Pub., 1994.

INDEX

A

Annualized turnaround cost (metrics), 86

Auditing, 64–66, 89: reliability, 64–66; compliance (metrics), 89

B

Backlog tracking, 29–30, 71: management system, 29–30

Benchmarking data, xix–xx, 1–14, 39, 41: performance comparison/ evaluation, 2–14; excuses, 4; identifying problems, 4–5; key drivers, 5; plant types, 5–14; pacesetter comparison/ evaluation, 5–14; routine maintenance performance, 5–6; turnaround maintenance performance, 7–11, 39, 41; duration and interval, 7–9; cost, 10–11; reliability performance, 12–14

Benchmarking, xi, xix–xx, 1–14, 39, 41: data, xix–xx, 1–14, 39, 41

Bibliography, 102

Business planning deliverables (turnaround maintenance), 46

C

Comparison/evaluation (benchmarking data), 5–14: routine maintenance, 5–6; turnaround maintenance, 7–11; duration and interval, 7–9; cost, 10–11; reliability, 12–14

Contractor counts (metrics), 70, 83, 85

Contractors, 70, 83, 85: counts, 70, 83, 85

Cost (maintenance), xviii–xix, 10–11, 21, 31, 71, 83: manpower, xviii, 31; sustainable, xix, 21; benchmarking, 10–11; metrics, 71, 83

Critical-path scheduling (turnaround maintenance), 40–41